Without Words
Mathematical Puzzles
to Confound and Delight

T0155216

James Tanton

TarquinGroup
www.tarquingroup.com

By the same author from Tarquin

More Without Words
35 more visual puzzles to test and delight you!
ISBN 978 1 907550 24 9
Ebook ISBN 978 1 858118 12 3

Publisher's Note:
We encourage you, before you start, to read the Introduction – these puzzles
will often test you and different types of puzzle will test you in different
ways. Some you may find easier than others – but if you get really stuck
there are some hints and tips on the last 3 pages.

To check answers and to find out more, go to www.tarquingroup.com and
search for Without Words - the link to get the answers and further reading is
on the relevant book's page.

© James Tanton 2015
ISBN 978 1 907550 23 2
Ebook ISBN 978 1 858118 11 6

tarquin publications
Suite 74, 17 Holywell Hill
St Albans, AL1 1DT, UK
info@tarquingroup.com
www.tarquingroup.com

A FEW BRIEF INFORMAL WORDS (!!)

WITHOUT WORDS is a collection of 36 immediately accessible but deeply mathematical puzzles, all designed to offer true joy in thinking mathematically in creative, innovative and surprising ways. Not a single word is written. These puzzles are universal: they transcend the barriers of language and culture, literally, and are thereby accessible to all people on this globe. Moreover, the puzzles themselves speak the universal truth of mathematics.

Many of these puzzles are hard and they are not intended to be solved in one sitting. (And sometimes the bulk of the challenge is figuring out what is being asked!) The idea is to sit with a puzzle or two over a number of days and wait for a moment of inspiration or insight to arrive. That is the true mathematical experience.

Mathematicians call such flashes of understanding *aha moments* and recognize that they cannot be called upon within minutes, or even hours, or even days. (Mathematicians don't believe that timed tests reflect the mathematical experience.) Moments of insight will come of their own accord and there is nothing to be done but let them take their time. There is never a need to rush in doing true mathematics!

For this reason, each puzzle should be mulled upon for quite some while. This book is designed to carry interest for many weeks, not just hours. Perhaps pull out a particular puzzle page and hang it as a poster on the wall. Let it become part of your subconscious. It's surprising how fruitful indirect mulling can be!

Some puzzles are relatively easy to answer, while others are quite deep. Some connect to current research problems in mathematics. Any relevant insight garnered from a particular puzzle should be deemed a success, even if it does not solve the challenge completely. (For those looking for guidance, a tips and techniques section appears at the end of the book.)

Mathematics can be compelling and frustrating, enlightening and humbling, but most of all it should be forever joyful. So on that note … enjoy!

JST

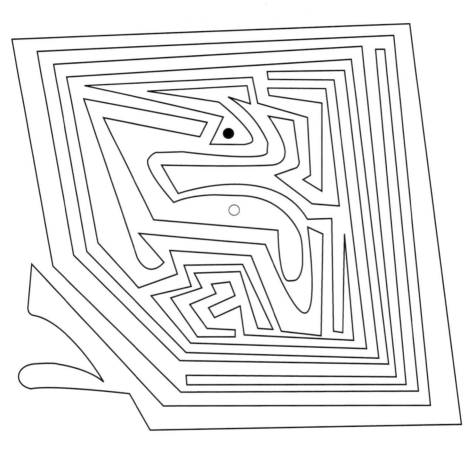

1

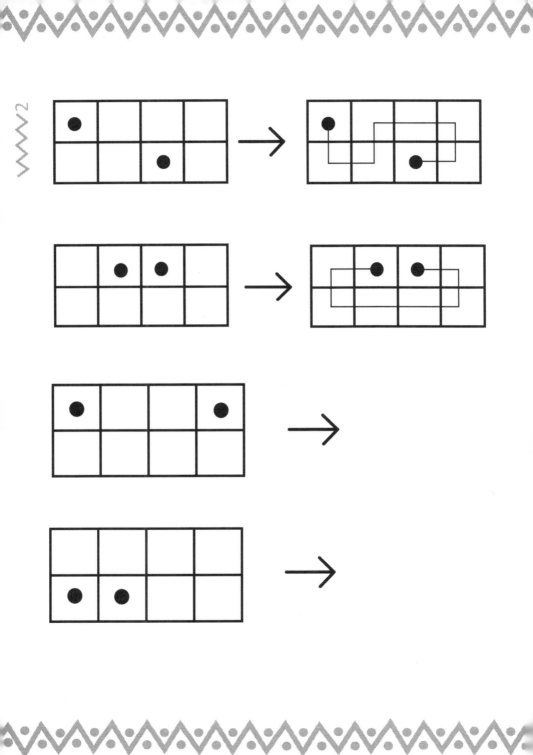

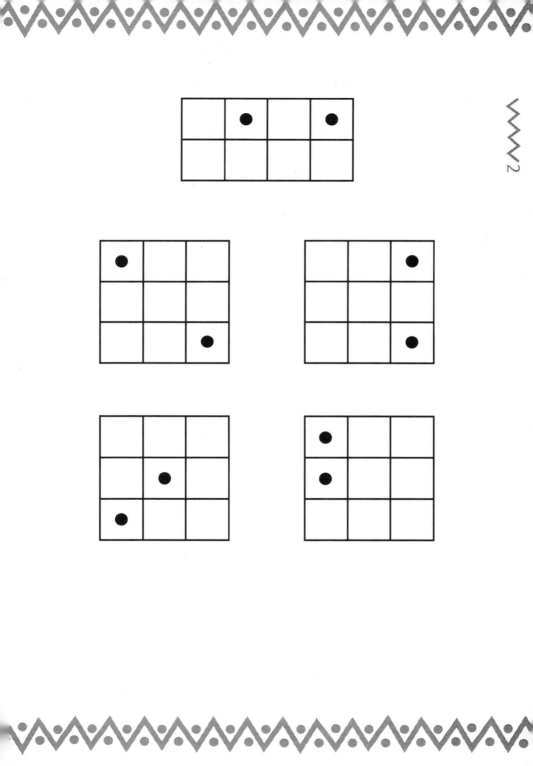

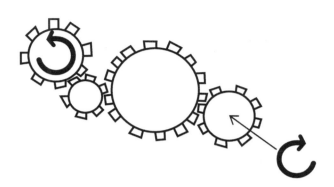

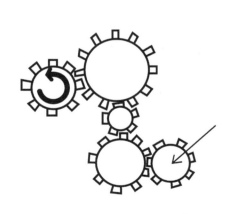

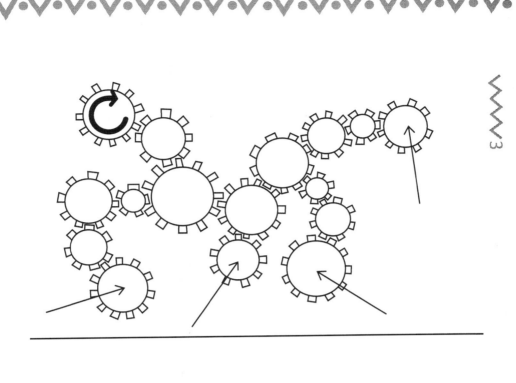

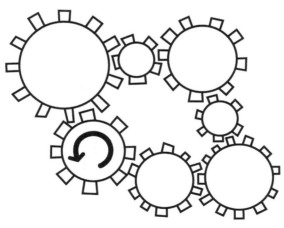

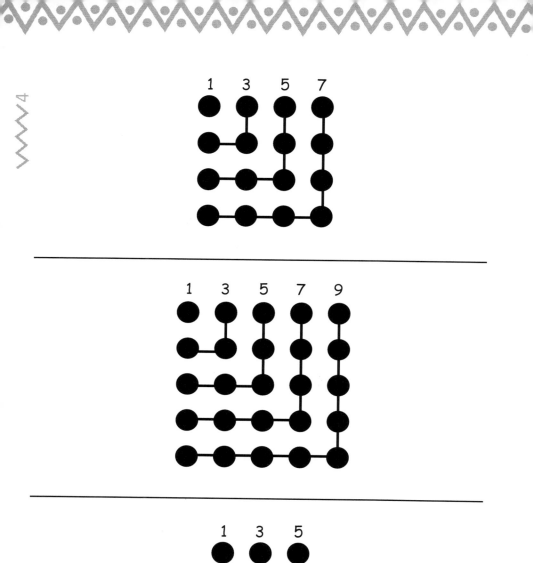

$$1 + 3 + 5 + 7 + 9 + 11 =$$

$$1 + 3 + 5 + 7 + 9 + 11 + 13 + 15 + 17 + 19 =$$

$$2 + 4 + 6 + 8 + 10 + 12 + 14 + 16 + 18 + 20 =$$

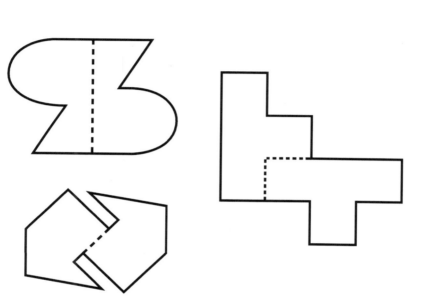

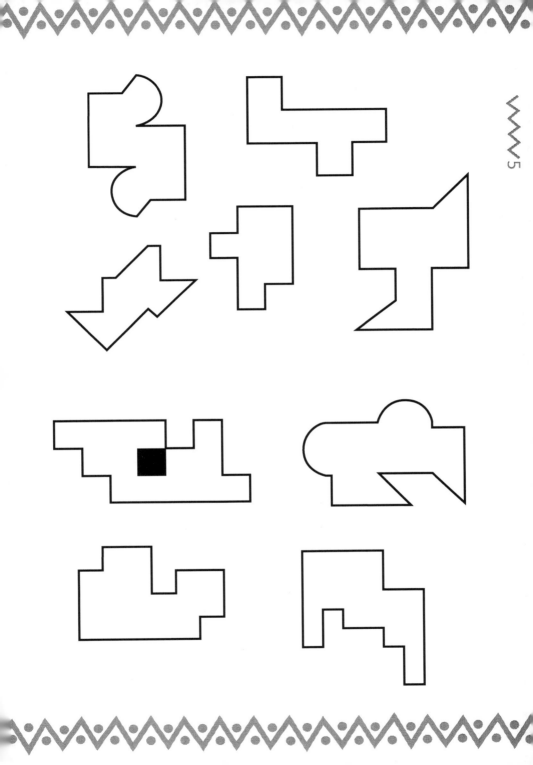

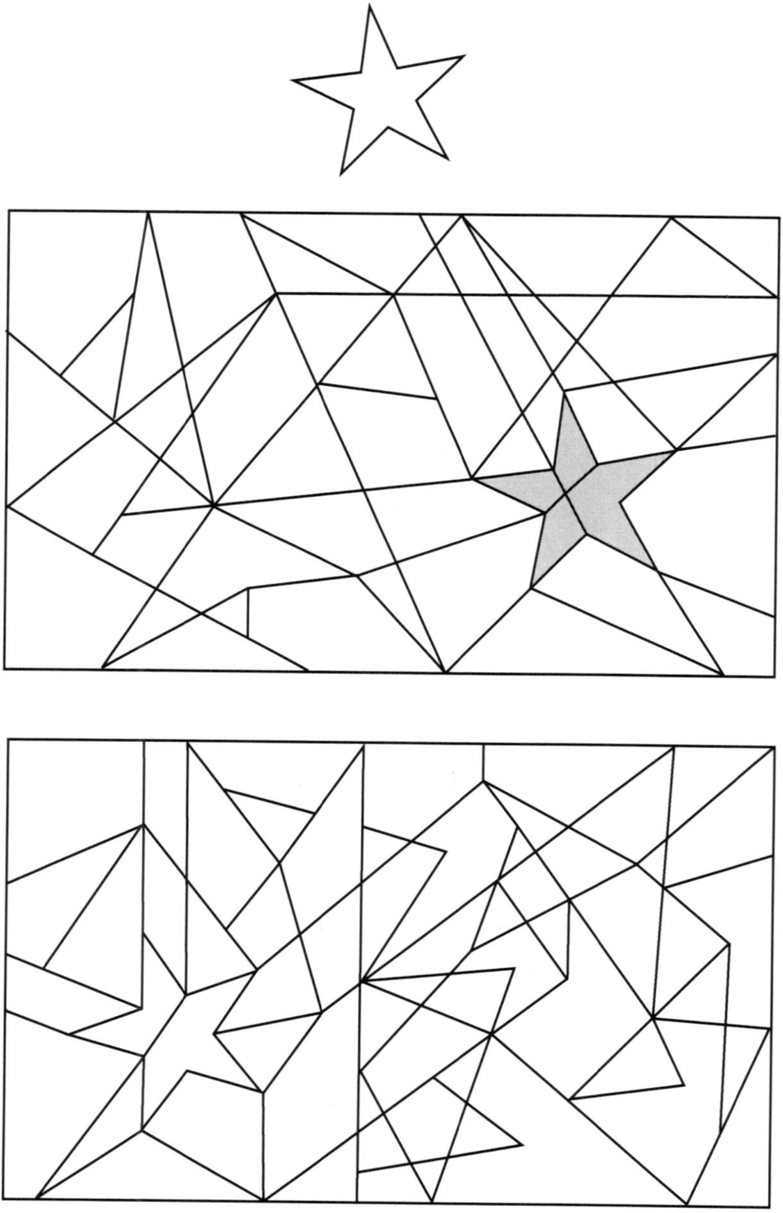

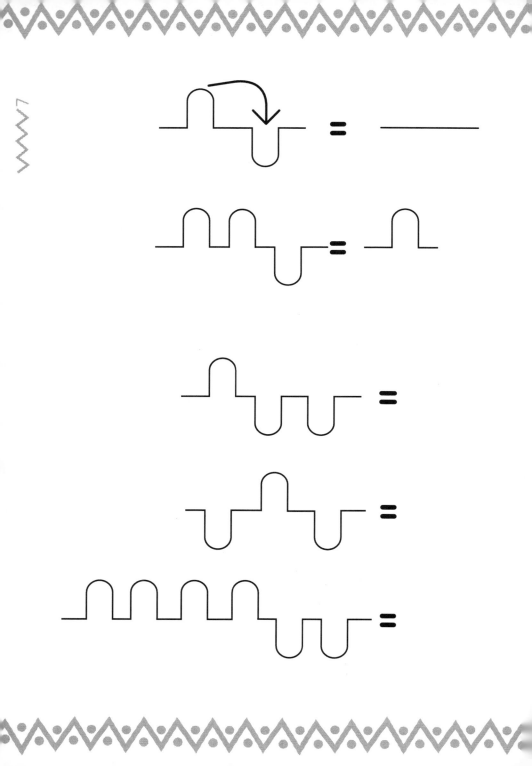

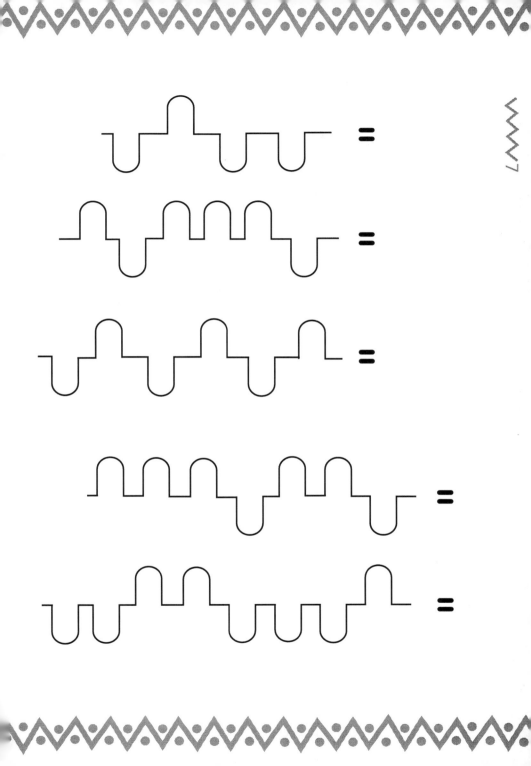

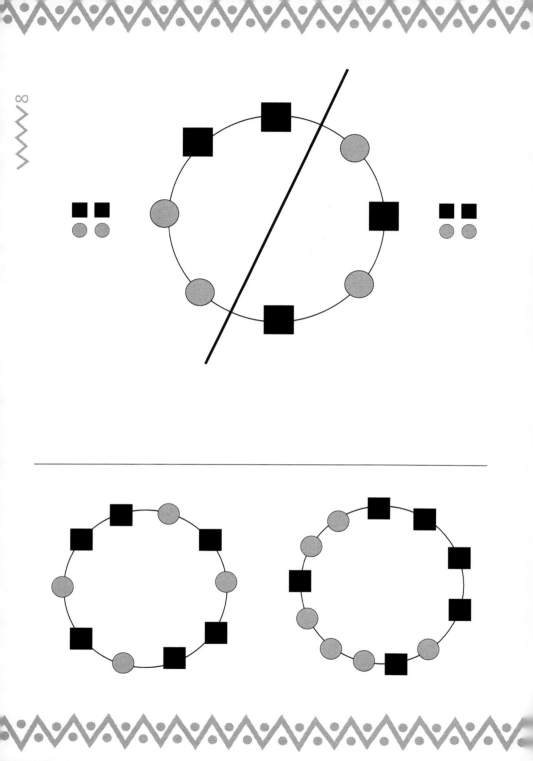

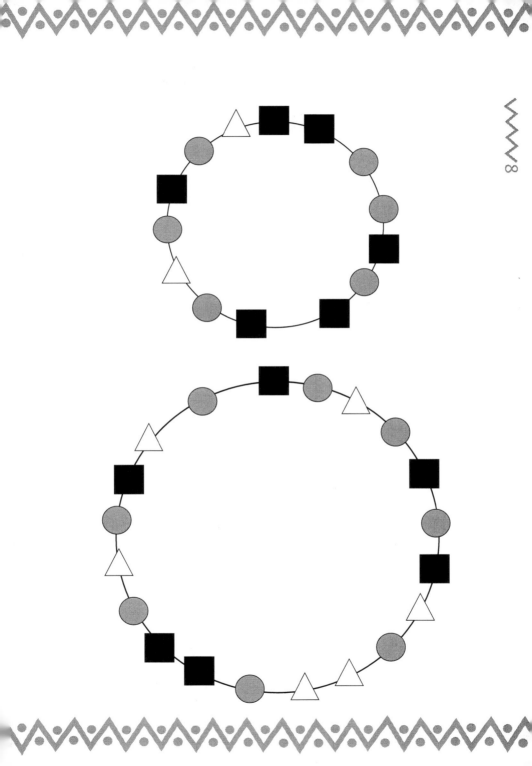

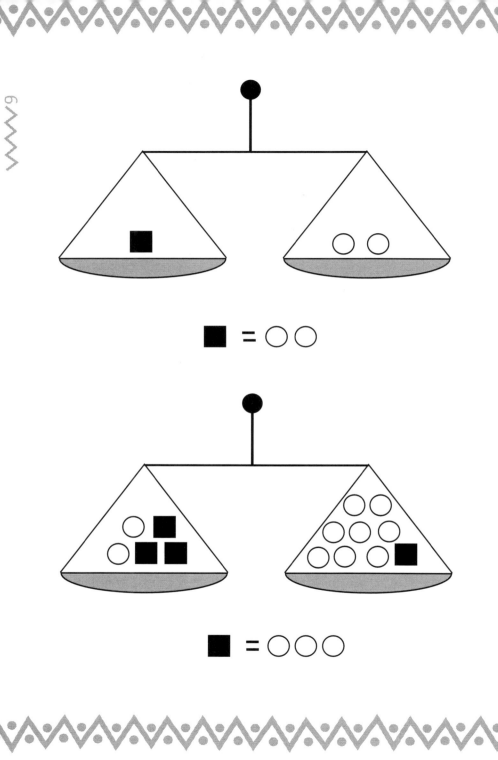

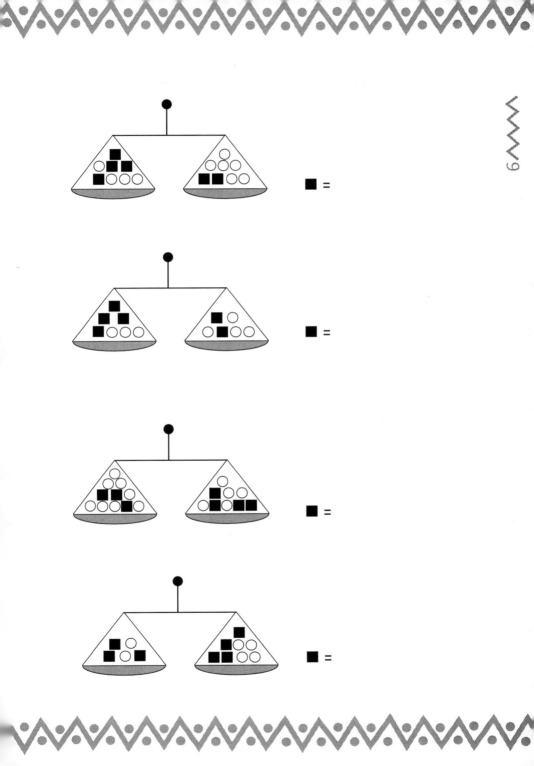

■ =

■ =

■ =

■ =

6

1

2

4

8

16

?

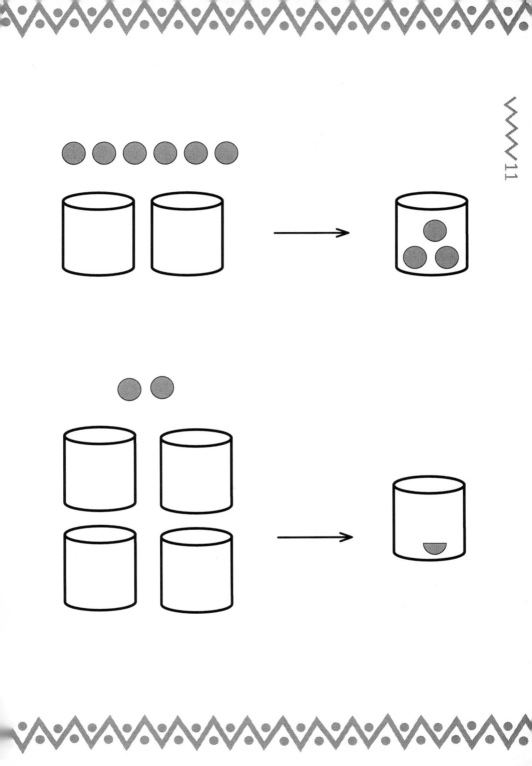

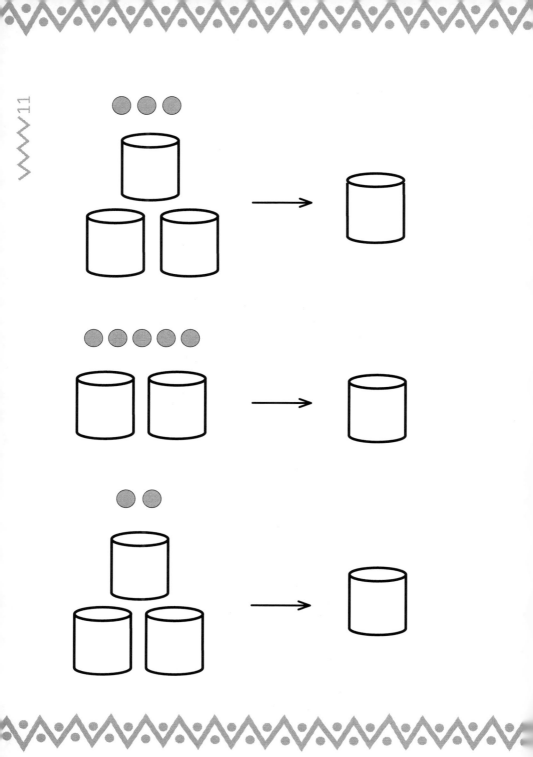

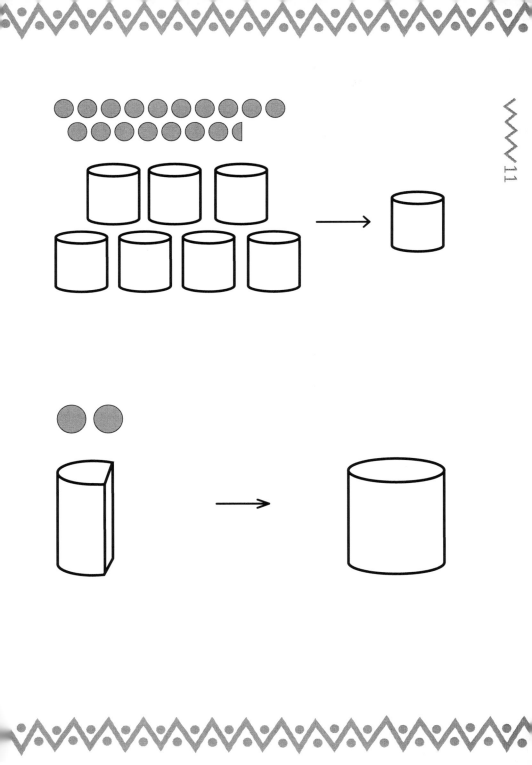

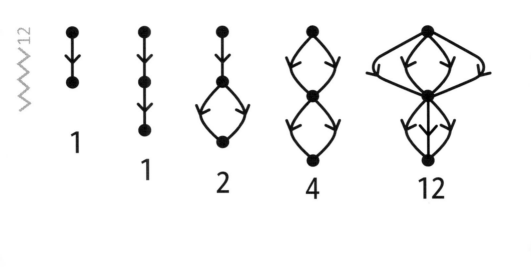

1   1   2   4   12

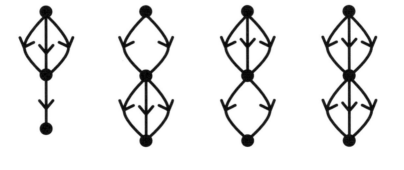

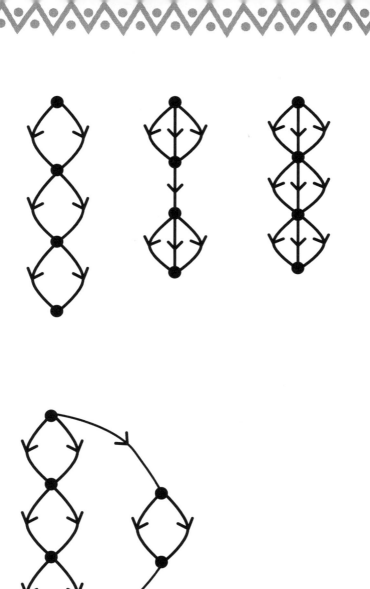

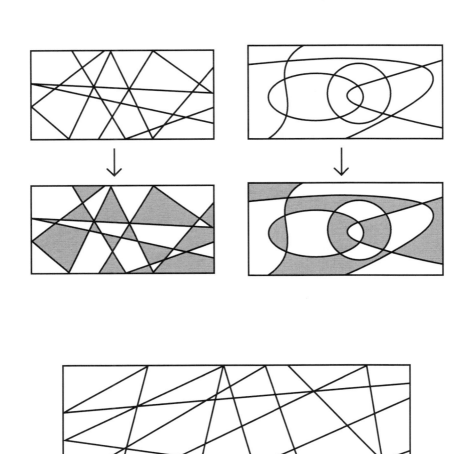

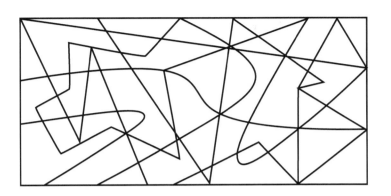

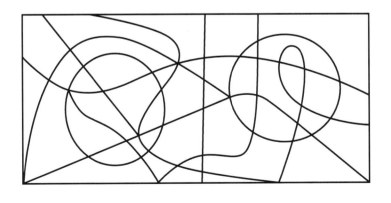

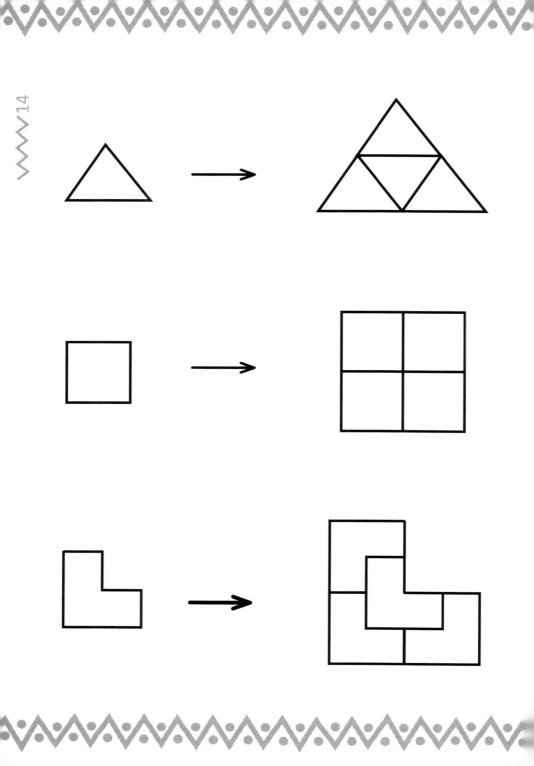

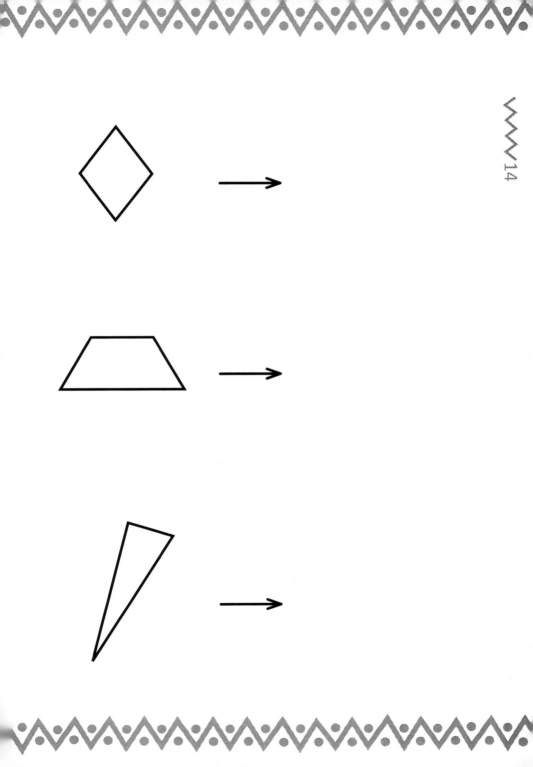

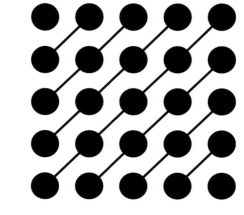

1 + 2 + 3 + 4 + 5 + 4 + 3 + 2 + 1

= 25

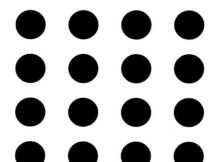

1 + 2 + 3 + 4 + 3 + 2 + 1

= 16

$$1 + 2 + 3 + 2 + 1$$
$$= \underline{\hspace{2cm}}$$

$$1 + 2 + 3 + 4 + 5 + 6 + 5 + 4 + 3 + 2 + 1$$
$$= \underline{\hspace{2cm}}$$

$$1 + 2 + 3 + 4 + 5 + 6 + 7 + 6 + 5 + 4 + 3 + 2 + 1$$
$$= \underline{\hspace{2cm}}$$

$$1 + 2 + 3 + \ldots + 99 + 100 + 99 + \ldots + 3 + 2 + 1$$
$$= \underline{\hspace{2cm}}$$

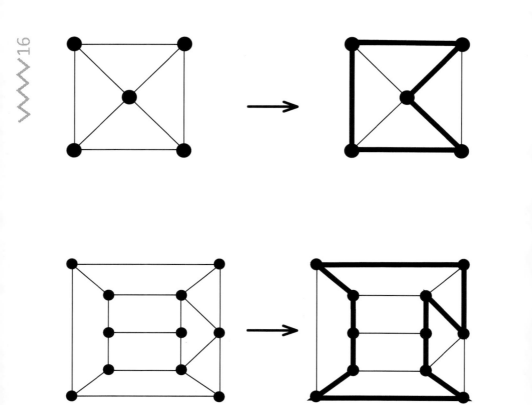

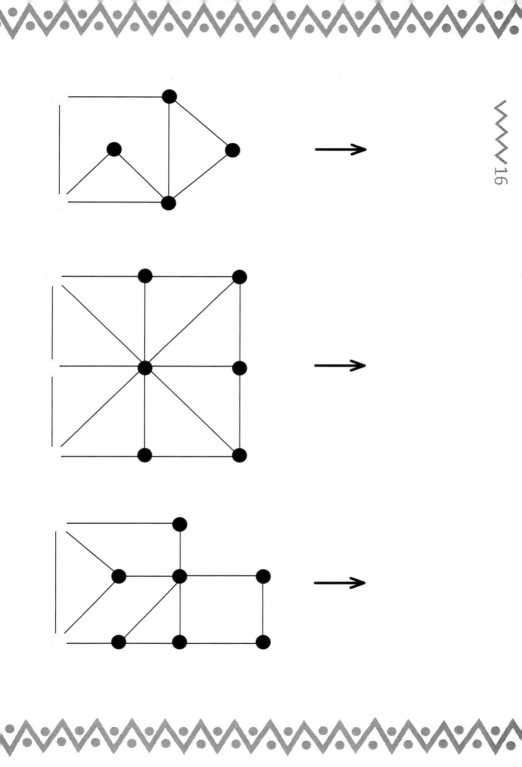

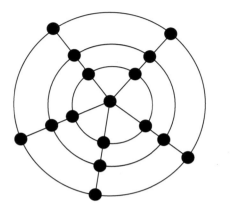

→

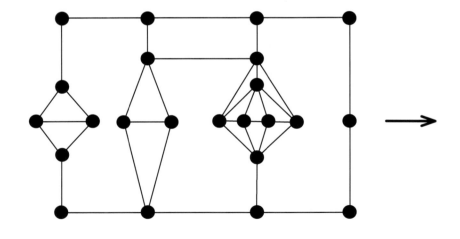

→

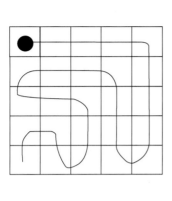
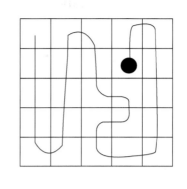

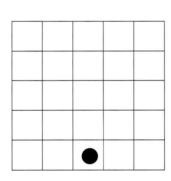
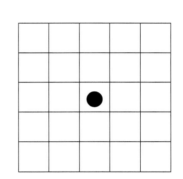

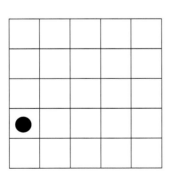
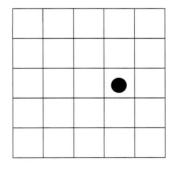

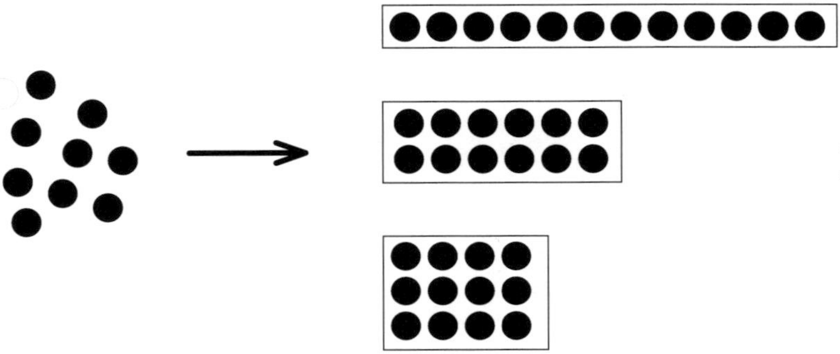

3

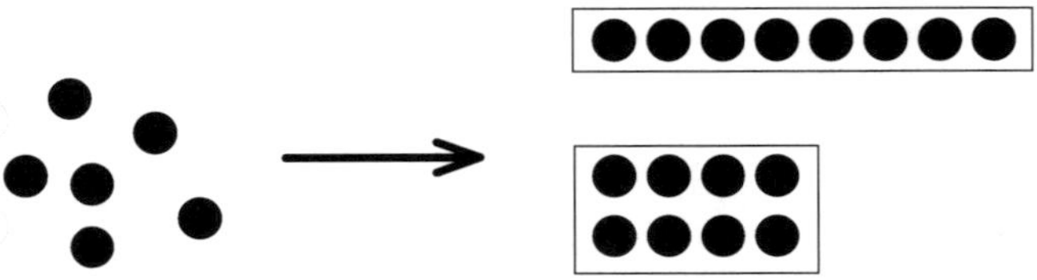

2

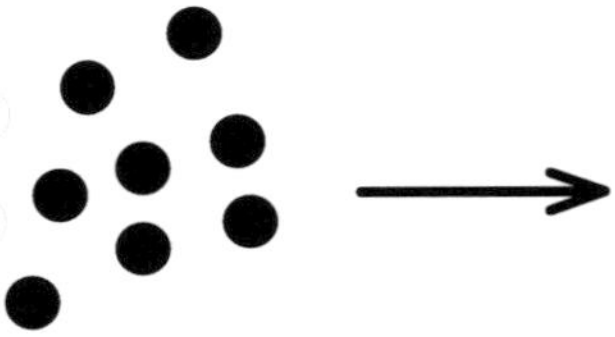

2

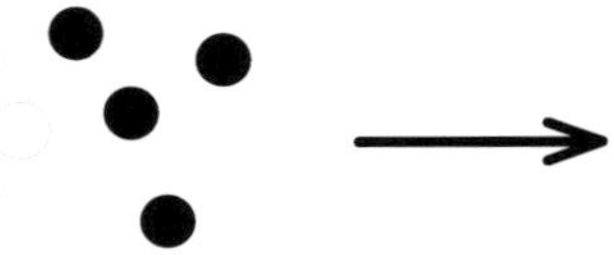

1

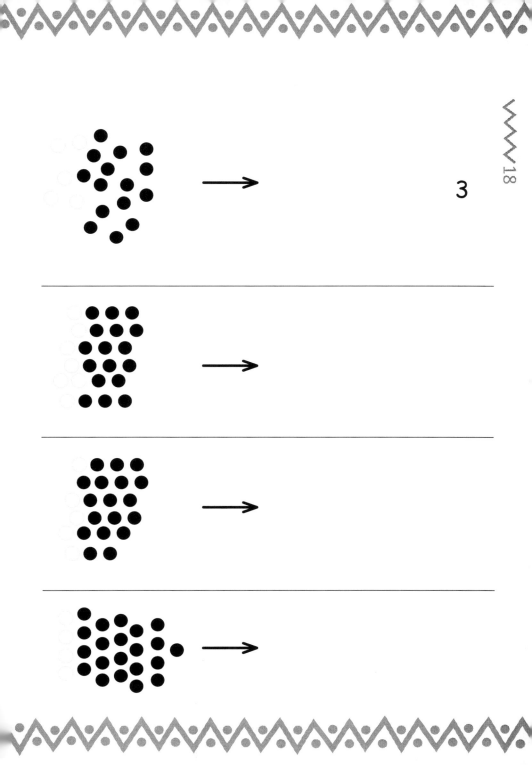

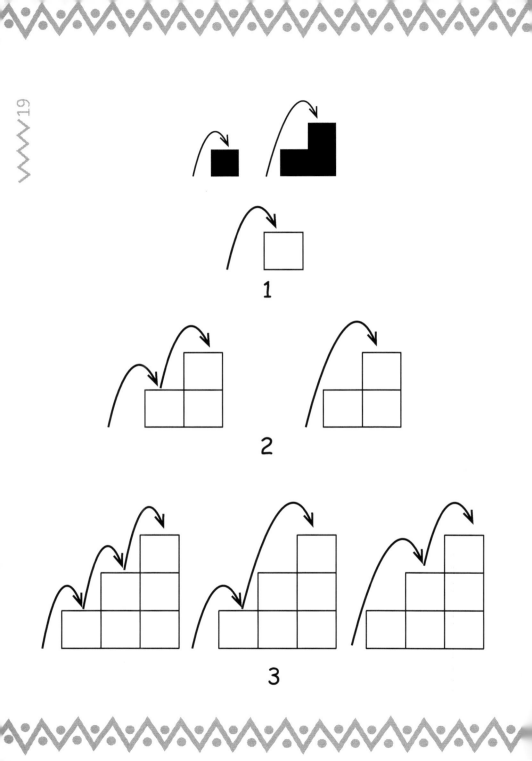

1

2

3

5

?

?

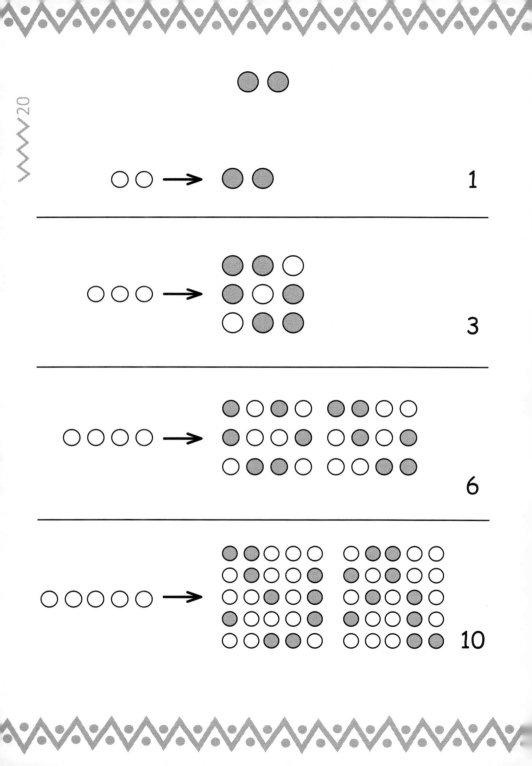

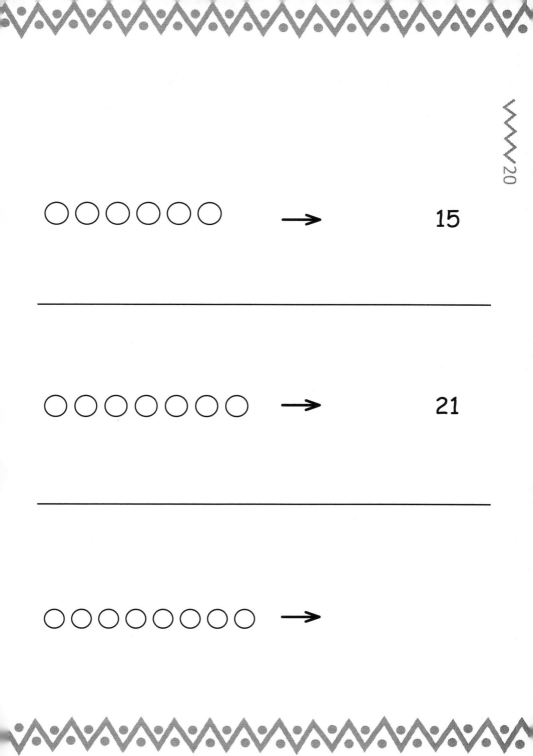

○○○○○○ → 15

○○○○○○○ → 21

○○○○○○○○ →

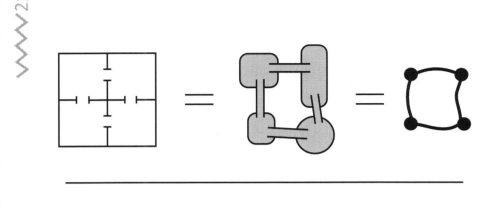

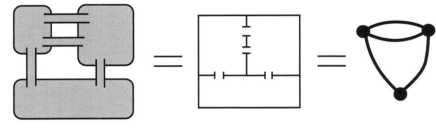

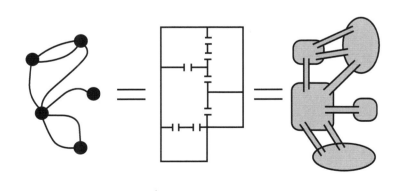

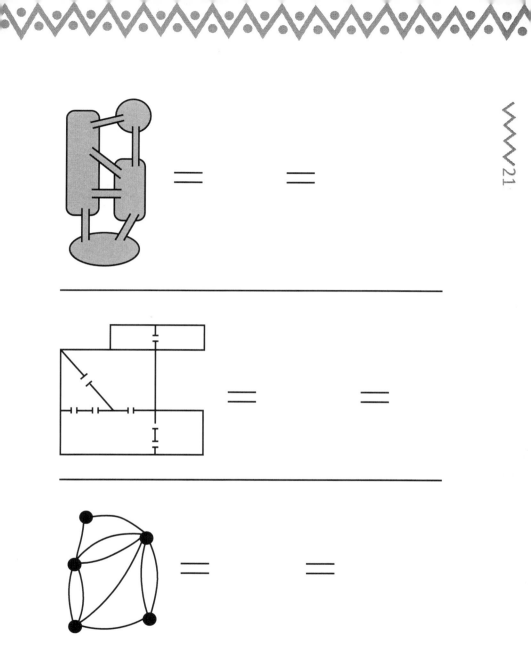

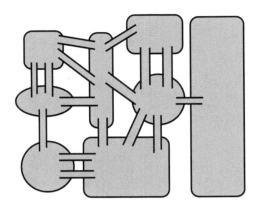

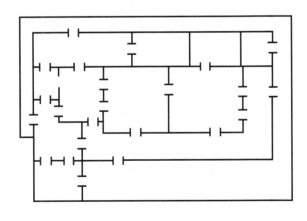

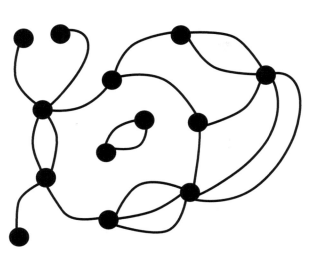

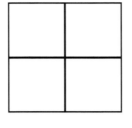

□ = 4   □□ = 2

□ = 2   ⊞ = 1
□

□ + □ + □□ + ⊞ = 9
4   2   2   1

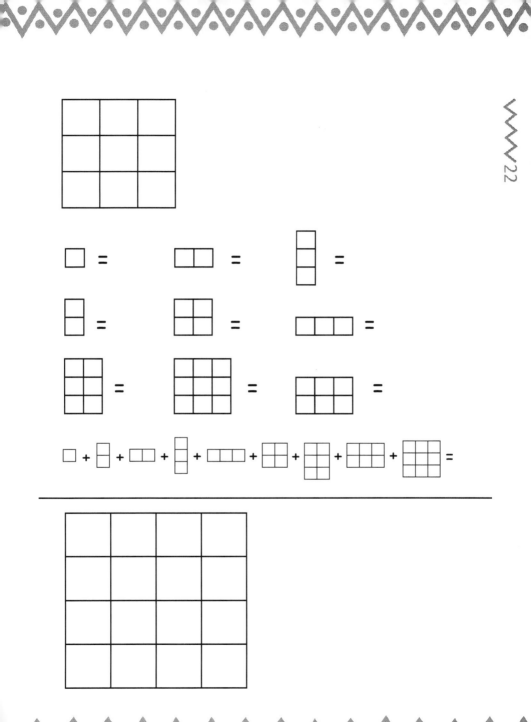

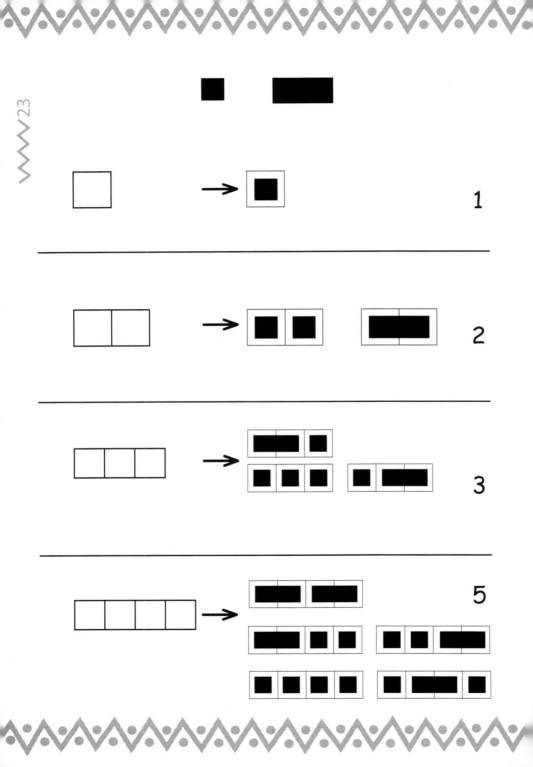

| | | | | | → | | 8 |

| | | | | | | → |

| | | | | | | | → |

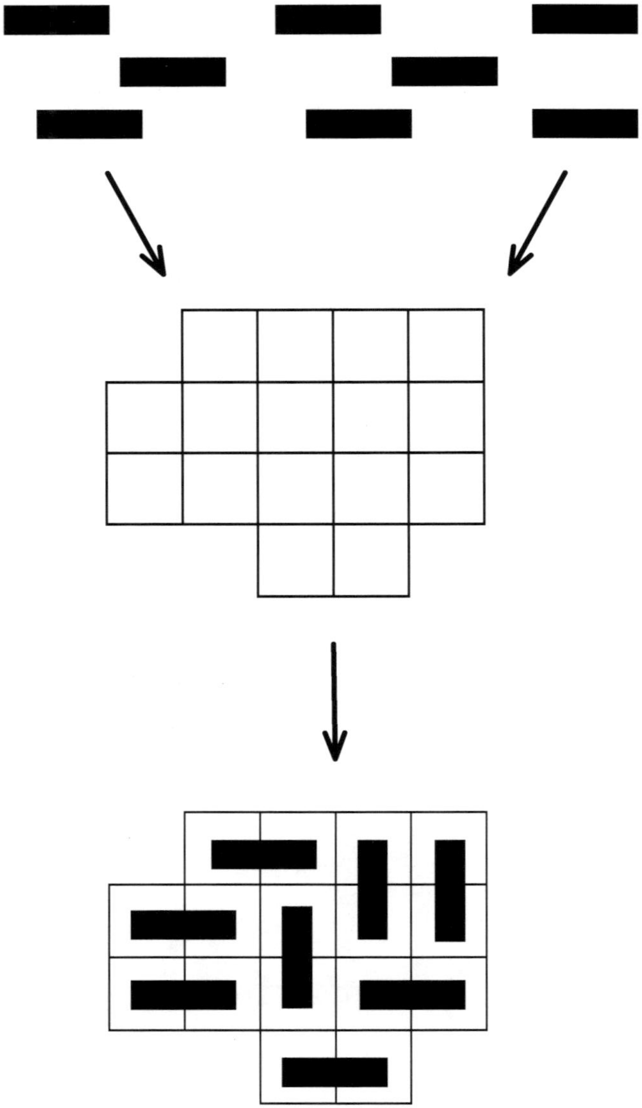

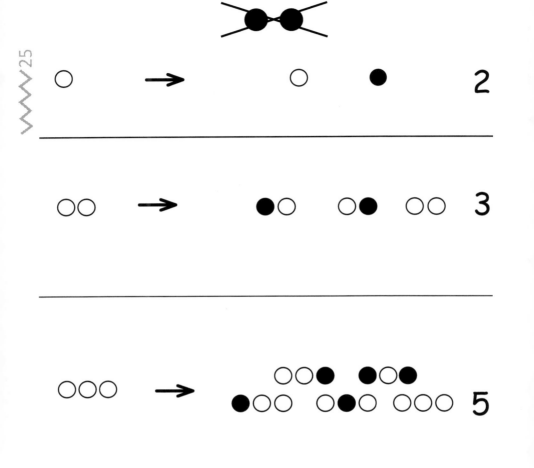

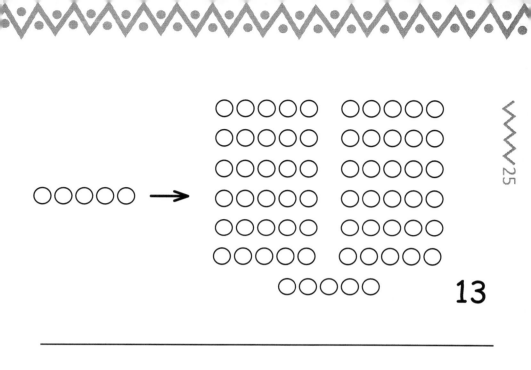

13

_____

 →

●● ●●●

● → ●● 1

●● → ●●● 1

●●●● → ●● | ●● 1

●●●● → ●●|●● ●●
●●● |●● 2

●●●●● → ●●● |●●●
●●|●● |●● 2

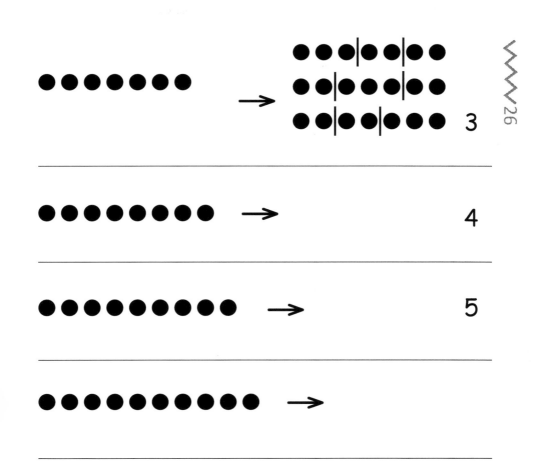

●●●●●● → ●●●|●●●|●● ●●●|●●●|●● ●●●|●●●|●● 3

●●●●●●●● → 4

●●●●●●●●● → 5

●●●●●●●●●● →

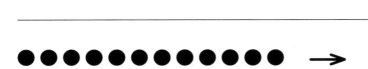

●●●●●●●●●●●● →

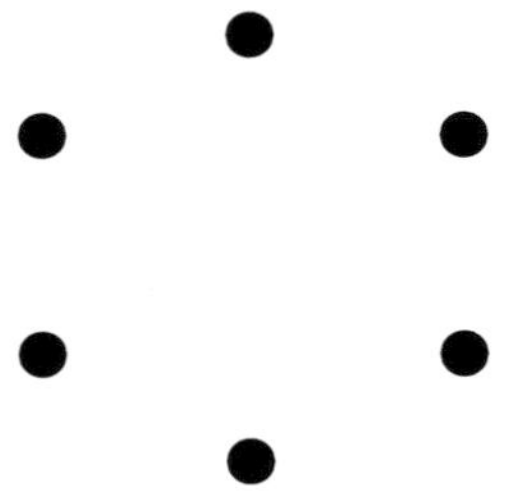

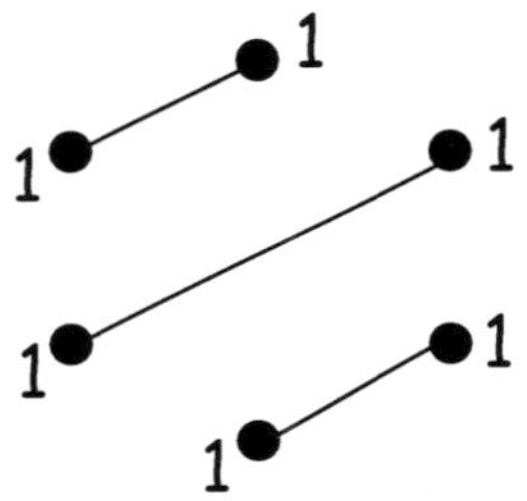
1
1
1
1
1
1

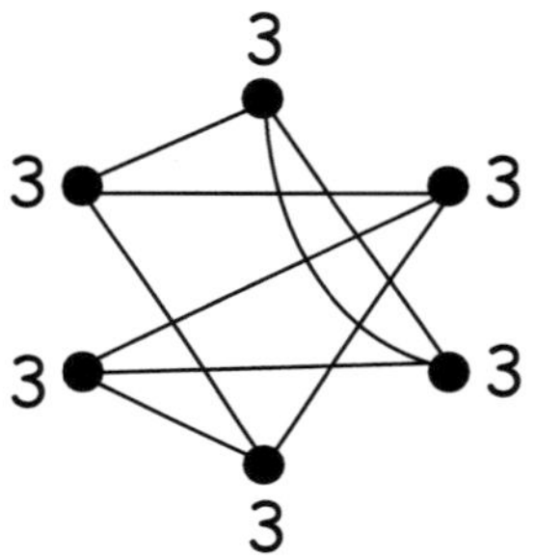
3
3
3
3
3
3

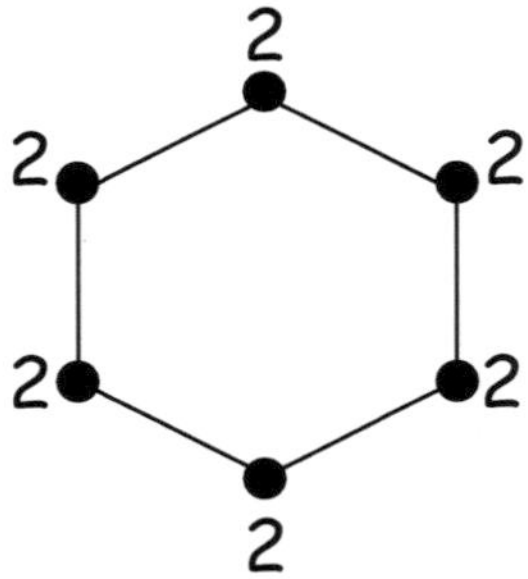
2
2
2
2
2
2

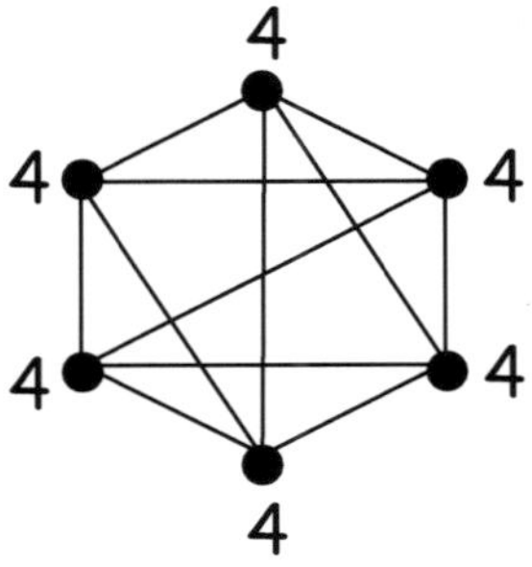
4
4
4
4
4
4

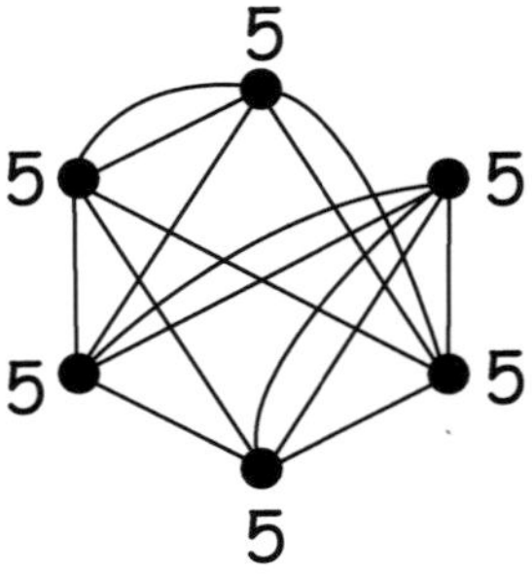
5
5
5
5
5
5

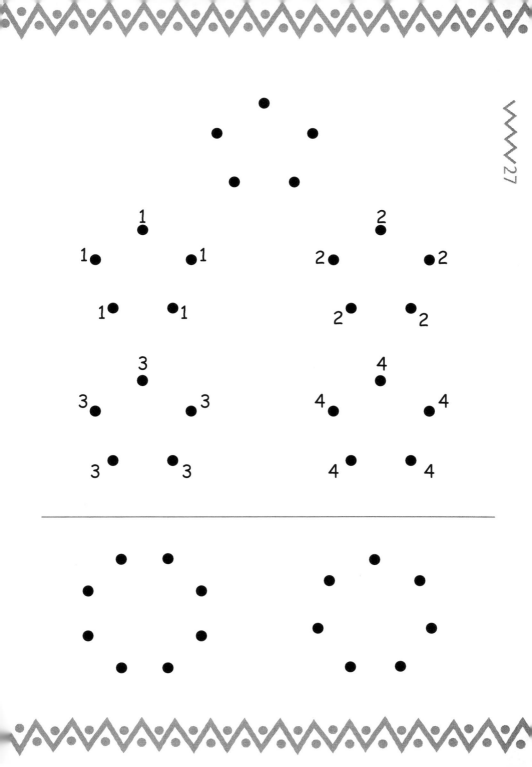

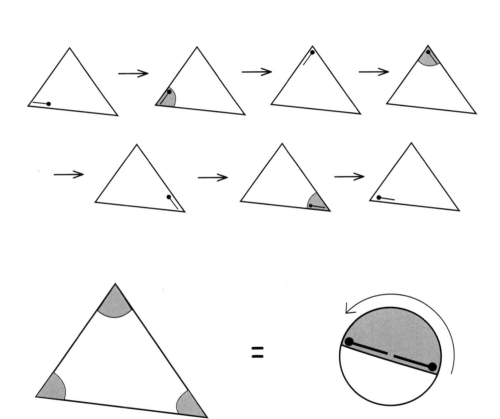

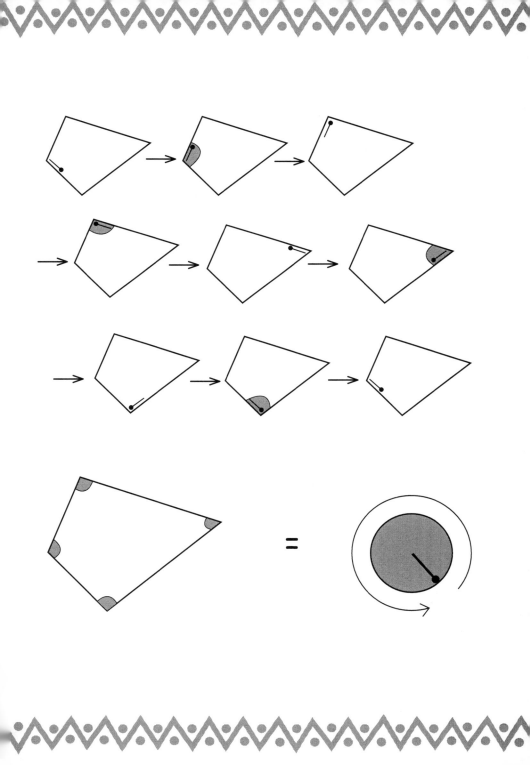

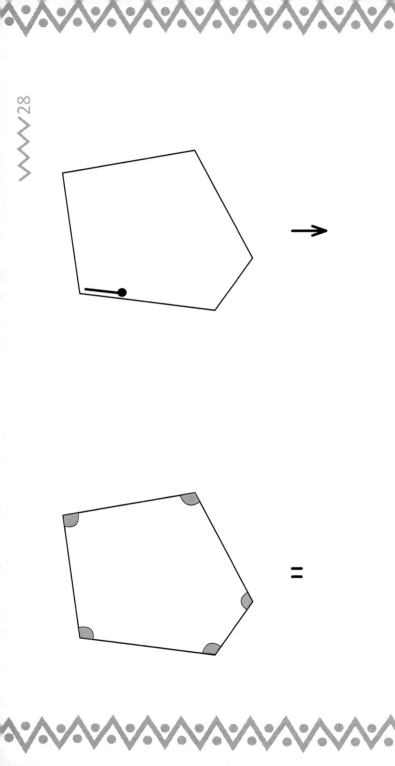

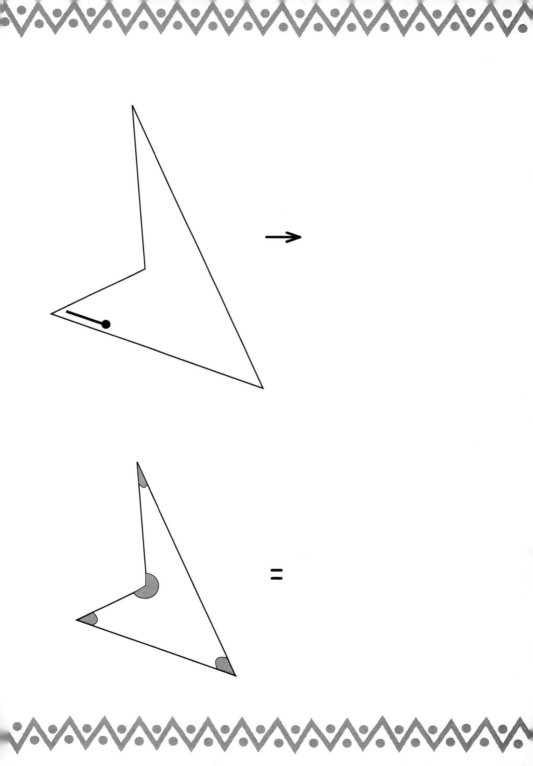

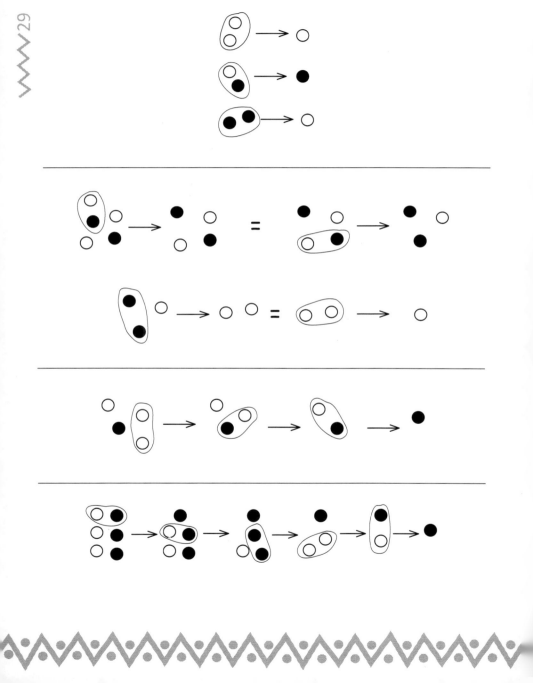

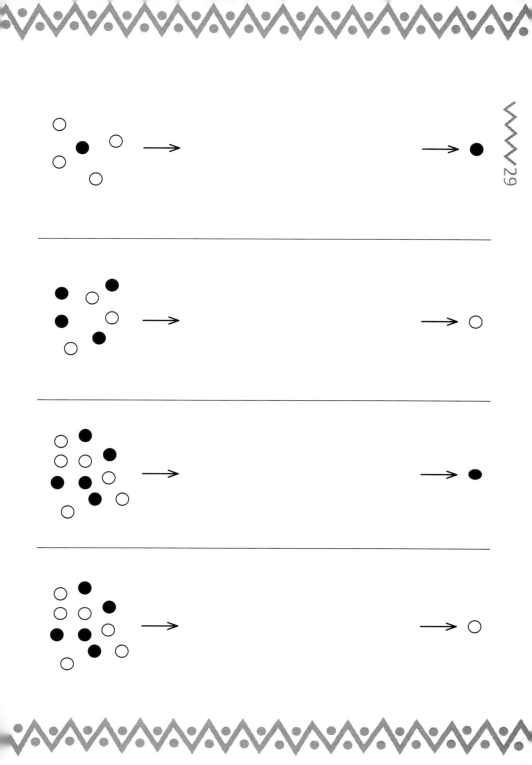

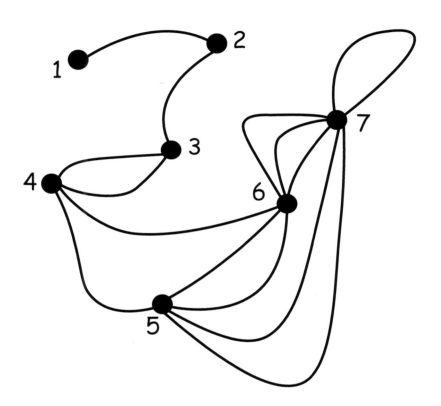

1 ●   ● 3

2 ●

                              ● 8

          4 ●

● 5

                    ● 7

      ● 6

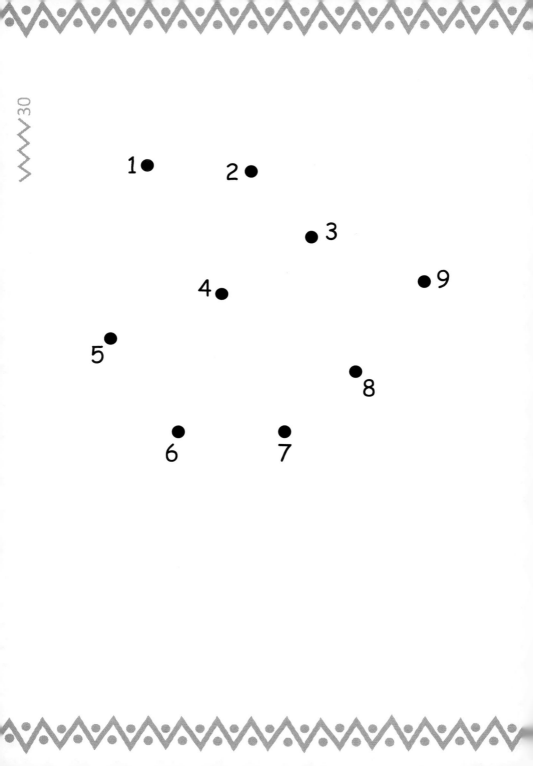

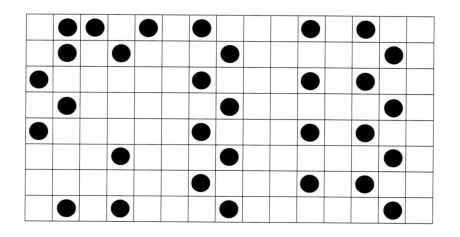

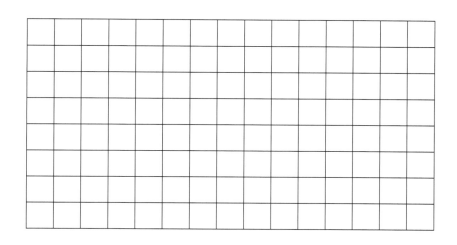

32

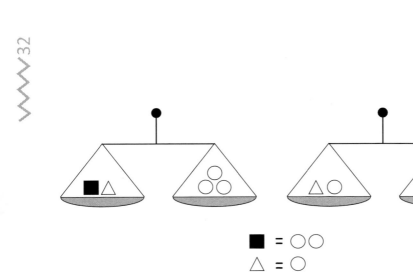

■ = ○○
△ = ○

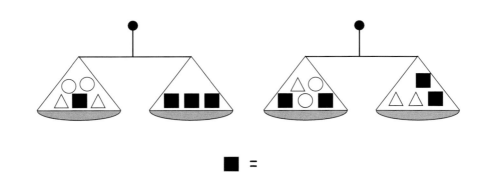

■ =
△ =

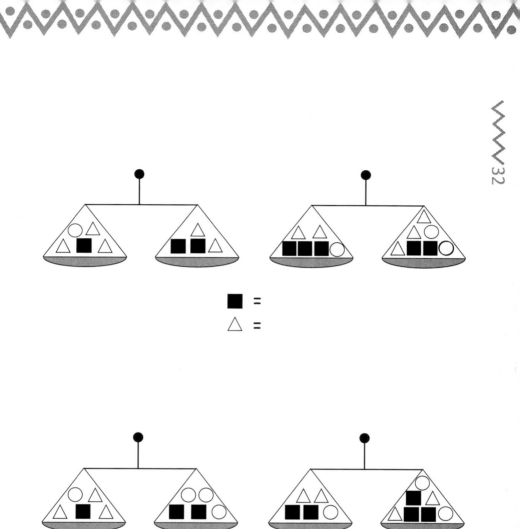

■ =
△ =

■ =
△ =

32

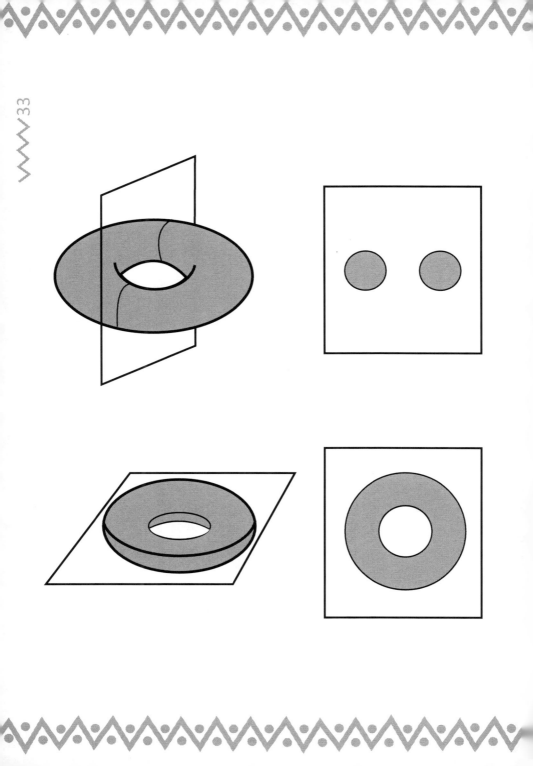

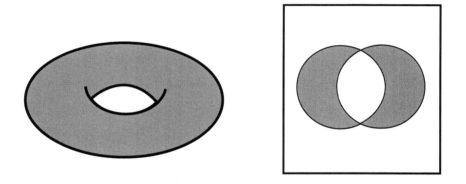

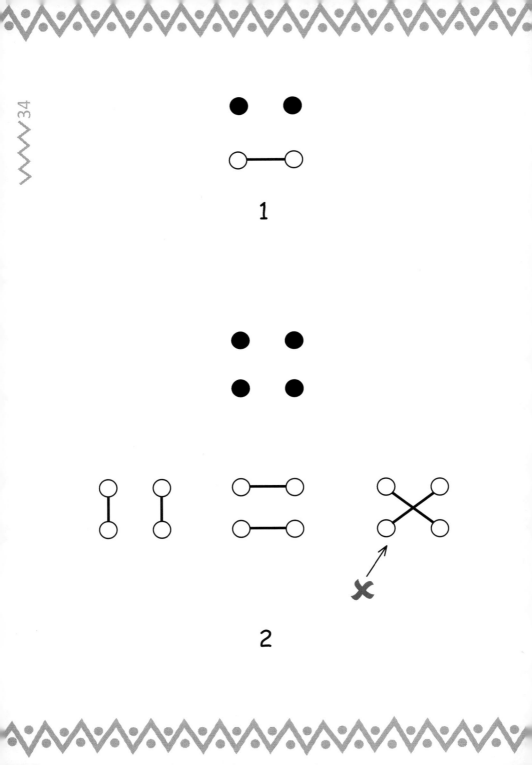

1

2

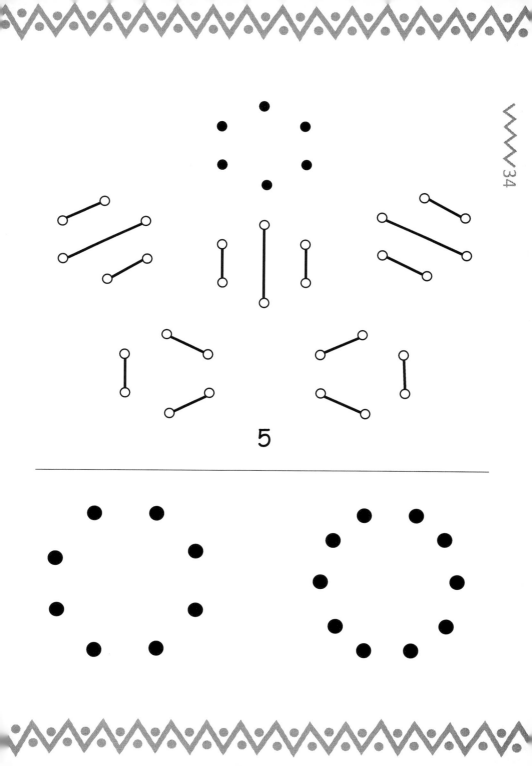

5

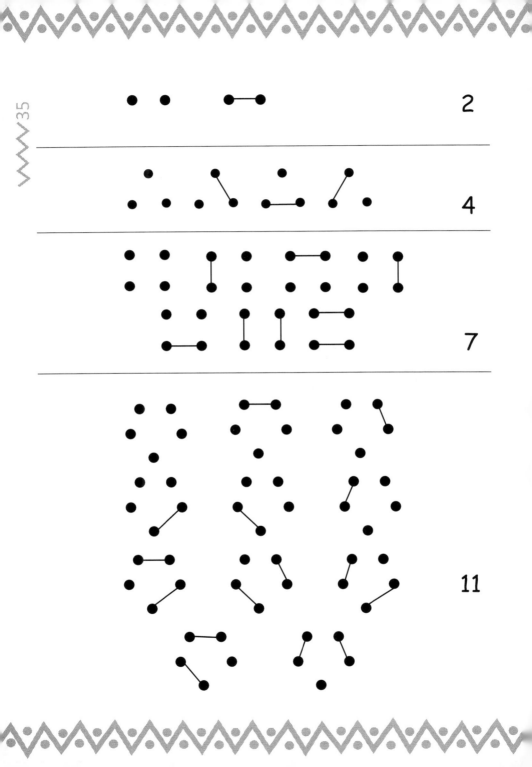

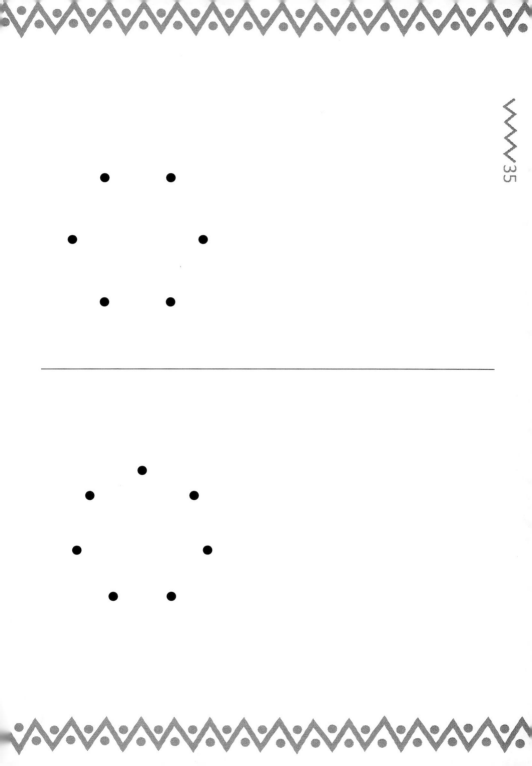

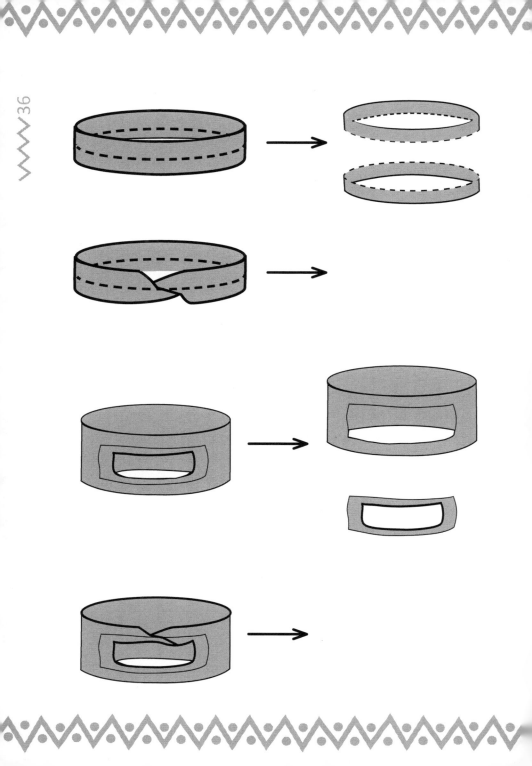

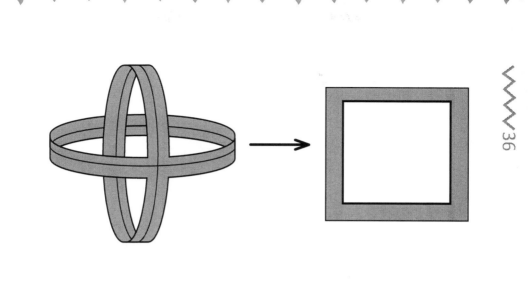

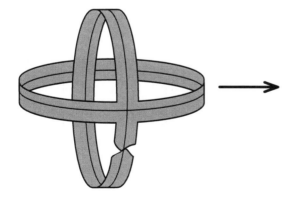

Techniques and Tips
... in case a few words are desired.

1. Is it always easy to identify the inside of a loop?

2. Can you walk vertical and horizontal steps from one dot to the other? Are puzzles like these always solvable? (And if not, why not?)

3. Counterclockwise, clockwise, counterclockwise, clockwise, counterclockwise, clockwise, …

4. The sum of the first four odd numbers is 16. The sum of the first five odd numbers is 25. Can you see what the sum of the first three odd numbers must be? Can you see the sum of the first six odd numbers? The first ten? How might you see the sum of the first ten even numbers?

5. Can you see how to divide each shape into two congruent pieces?

6. Mathematics is all about looking for structure in apparent chaos. This puzzle is a metaphor. Can you find the star?

7. This puzzle is playing with piles and holes in a sandbox. (Take this to thinking about the arithmetic of positive and negative integers.)

8. Can you always divide a necklace into two pieces so that each half has equal counts of beads of each type?

9. In terms of circles, what must be the weight of each square to make the scales balance?

10. If a line is drawn between each and every pair of dots amongst six dots on the rim of a circle, into how many pieces will the circle be divided? Can we trust patterns? Warning: There are actually two answers to this puzzle depending on whether you like to place dots symmetrically or don't mind lopsidedness. (Why the difference?)

11. If oranges are shared equally among containers, how many oranges per whole container will we get?

12. Following the one-way roads, how many different routes are there from top to bottom?

13. Can "maps" always be coloured with just two colours? (Any two regions sharing a length of border must be assigned opposite colours.) If something can go wrong, what, exactly, is the obstacle?

14. Can four copies of each figure stack to make a larger copy of itself?

15. Look at the diagonals of a square array of dots and discover an amazing arithmetic fact! (I bet you can now add together all the numbers from 1 up to 100 and back down again in a flash of a second.)

16. Can you walk a loop along the lines that visits each and every dot exactly once?

17. Taking only vertical and horizontal steps walk a journey from the indicated dot to visit each and every cell exactly once.

18. How many rectangles can you make? (Let's deem ninety-degree rotations as equivalent rectangles.) Which numbers are resistant to making meaningful rectangles?

19. Taking only one or two steps at a time, in how many different ways can one climb a short staircase?

20. Given a row of white dots, in how many ways can one select two to be coloured grey?

21. Rooms, islands and bridges, and dots and lines. Can you use all three ways to represent the same configuration?

22. One can find a total of 9 rectangles (including squares) in a two-by-two grid of squares. How many rectangles, in total, can you find in larger grids of squares? (Is there are way to use WW20 to help count them?)

23. Given 1-by-1 and a 1-by-2 tiles, in how many ways can one tile a row of squares? (Does this feel like WW19?)

24. Using 1-by-2 and 2-by-1 tiles (dominos), can you completely tile the given plans? If not, can you succinctly describe the general obstacle?

25. Given a row of dots, count how many ways to colour some of them black avoiding neighbouring black dots.

26. Count how many ways to break a row of dots into groups of two and/or three.

27. Draw lines between dots so that each dot has the indicated number of lines emanating from it. Can it always be done? (If not, what exactly is the obstacle?)

28. Sliding a pencil inside a triangle shows that the three angles in the triangle sum to half a turn. Sliding a pencil in a (convex) quadrilateral shows that the four angles of the figure sum to a full turn. Explore sums of angles in other polygons this way.

29. Can you always win this game of solitaire? Start with a collection of black and white dots. A "move" consists of erasing any two dots of your choice and replacing them with a single dot according to the rules:

    If you erase two identical dots, replace them with a white dot.

    If you erase two dots of opposite colours, replace them with a black dot.

    As you play this game the count of dots decreases. Can you end the game with a single dot of the colour indicated? (Does it matter what strategy you follow as you play the game? Might these games be "rigged"?)

30. Like WW27, draw lines between dots so that each dot has the indicated number of lines emanating from it.

31. Try numbering the cells left to right, top to bottom, 1,2, 3, 4, 5, …., 120. What's special about the placement of the dots? (See also WW18).Can you continue the

pattern of dots? Are certain columns of the grid certain to be empty? Any other observations?

32. Like WW9, in terms of circles what must be the weight of each square and each triangle to make the scales balance?

33. Is there a third way to slice a donut to see two perfect circles on the slicing plane?

34. In how many ways can an even count of people sitting around a table simultaneously reach across to shake hands? (Assume no one's arms cross.)

35. In how many ways can some, none, or all people sitting around a table reach across to shake hands? (Again assume no one's arms cross.)

36. What happens if you cut these famous objects along the central lines indicated? (Care to make these constructions out of paper and physically do the cutting?)

There are solutions to the puzzles - and a wealth of other wonderful mathematical material – on the author's website: **www.jamestanton.com**.

More from Tarquin

If you've enjoyed this book then you'll enjoy its sister book More Without Words, too – and you might be interested in the Without Words Poster Set that accompanies the books – full details and images on www.tarquingroup.com.

There you will also find puzzles, books, dice and posters all with a mathematical theme. Our products are used worldwide by teachers, tutors, parents and recreational mathematicians.

Tarquin has been publishing for more than 35 years and we have something for everyone!

www.tarquingroup.com

For a full catalogue email us at info@tarquingroup.com
or write to us at:
Tarquin
Suite 74, 17 Holywell Hill
St Albans
AL1 1DT
United Kingdom
Our books and many other products are also available on Amazon and many other retailers worldwide.